EXPERIMENTS WITH A HAND LENS

A TRUE BOOK

by

Salvatore Tocci

Children's Press®
A Division of Scholastic Inc.
New York Toronto London Auckland Sydney
Mexico City New Delhi Hong Kong
Danbury, Connecticut

Reading Consultant
Nanci R. Vargus, Ed.D
Primary Multiage Teacher
Decatur Township Schools
Indianapolis, Indiana

Science Consultant
Robert Gardner

The author and publisher are not responsible for injuries or accidents that occur during or from any experiments. Experiments should be conducted in the presence of or with the help of an adult. Any instructions of the experiments that require the use of sharp, hot, or other unsafe items should be conducted by or with the help of an adult.

Library of Congress Cataloging-in-Publication Data

Tocci, Salvatore.
Experiments with a hand lens / Salvatore Tocci
p. cm. – (A True book)
Includes index.
ISBN 0-516-22506-5 (lib. bdg.) 0-516-26994-1 (pbk.)
1. Lenses—Experiments—Juvenile literature. [1. Lenses—Experiments. 2. Magnifying glasses—Experiments. 3. Experiments.] I. Title II. Series.

QC385.5.T63 2002
507'.8—dc21 00-049987

1 2 3 4 5 6 7 8 9 10 R 11 10 09 08 07 06 05 04 03 02

Contents

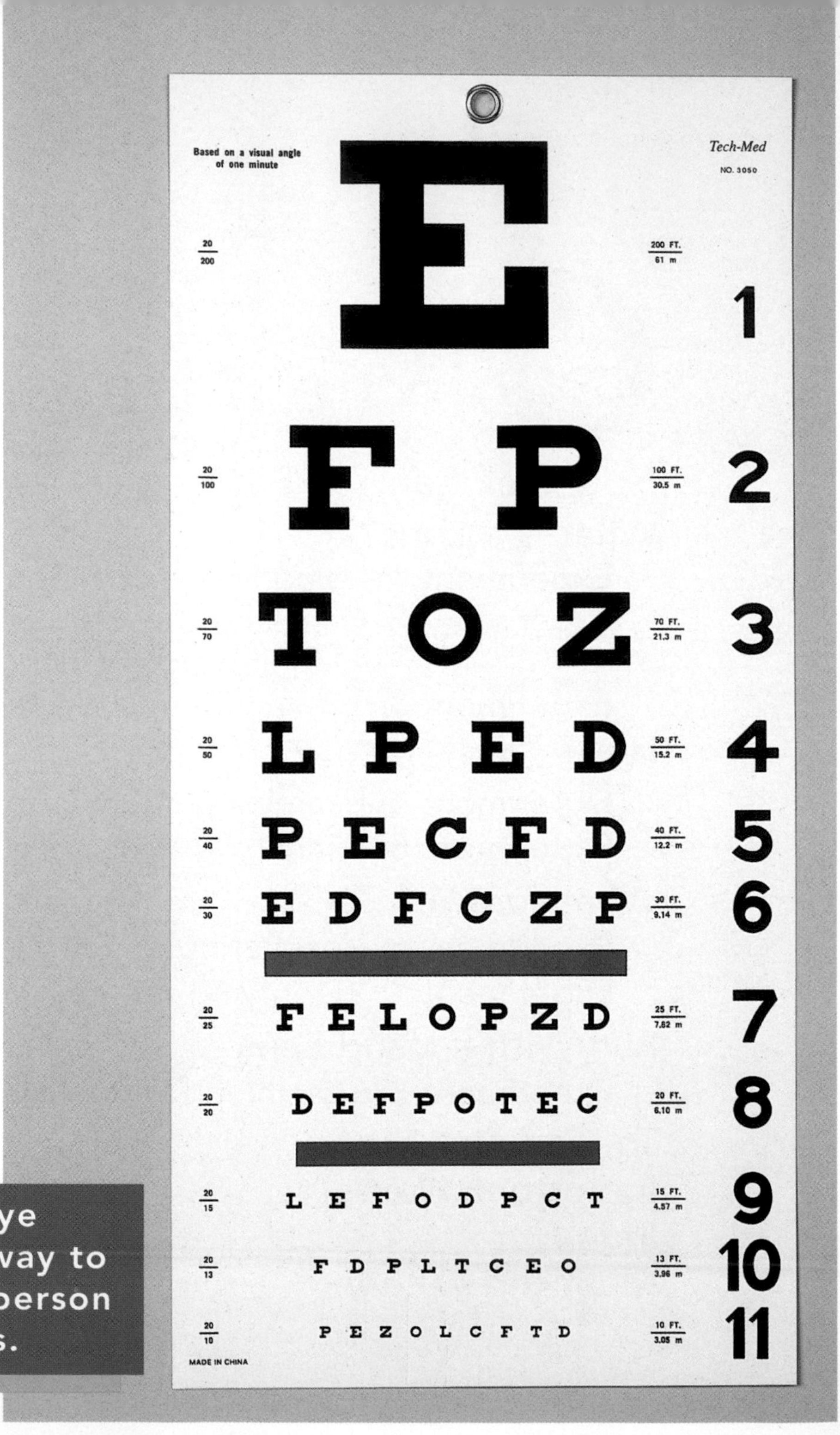

Reading an eye chart is one way to find out if a person needs glasses.

How Well Can You See?

Have you ever had your eyes examined? You probably have had your eyes checked to find out how well you can see. You may have been asked to read the letters on a chart from a distance of 20 feet (6 meters). As you read down the chart, the letters got smaller and smaller.

If you were able to read the line of letters marked 20, you have normal vision. Your vision is said to be 20/20 because you were able to read down to the line marked 20 from a distance of 20 ft.

Many people cannot read the letters on the line marked 20. For example, they may be able to read the letters only down to the line marked 40. Such people have 20/40 vision. A person with 20/40 vision cannot see as well as a person with 20/20 vision.

A person with 20/40 vision can drive a car without eyeglasses in all fifty states.

Some people have 20/10 vision. A person with this vision sees even better than a person with 20/20 vision. Even a person with 20/10 vision, however, cannot see everything well. For example, many objects are too small to be seen with just the eyes. These small objects can be seen only with the help of a lens.

What Is a Lens?

To see something, you need light. To see something well, the light rays that enter your eyes must be focused in just the right way. Focusing light rays is the job of the lens, which is at the front of the eye. The lens bends light rays as they pass through it. In a person

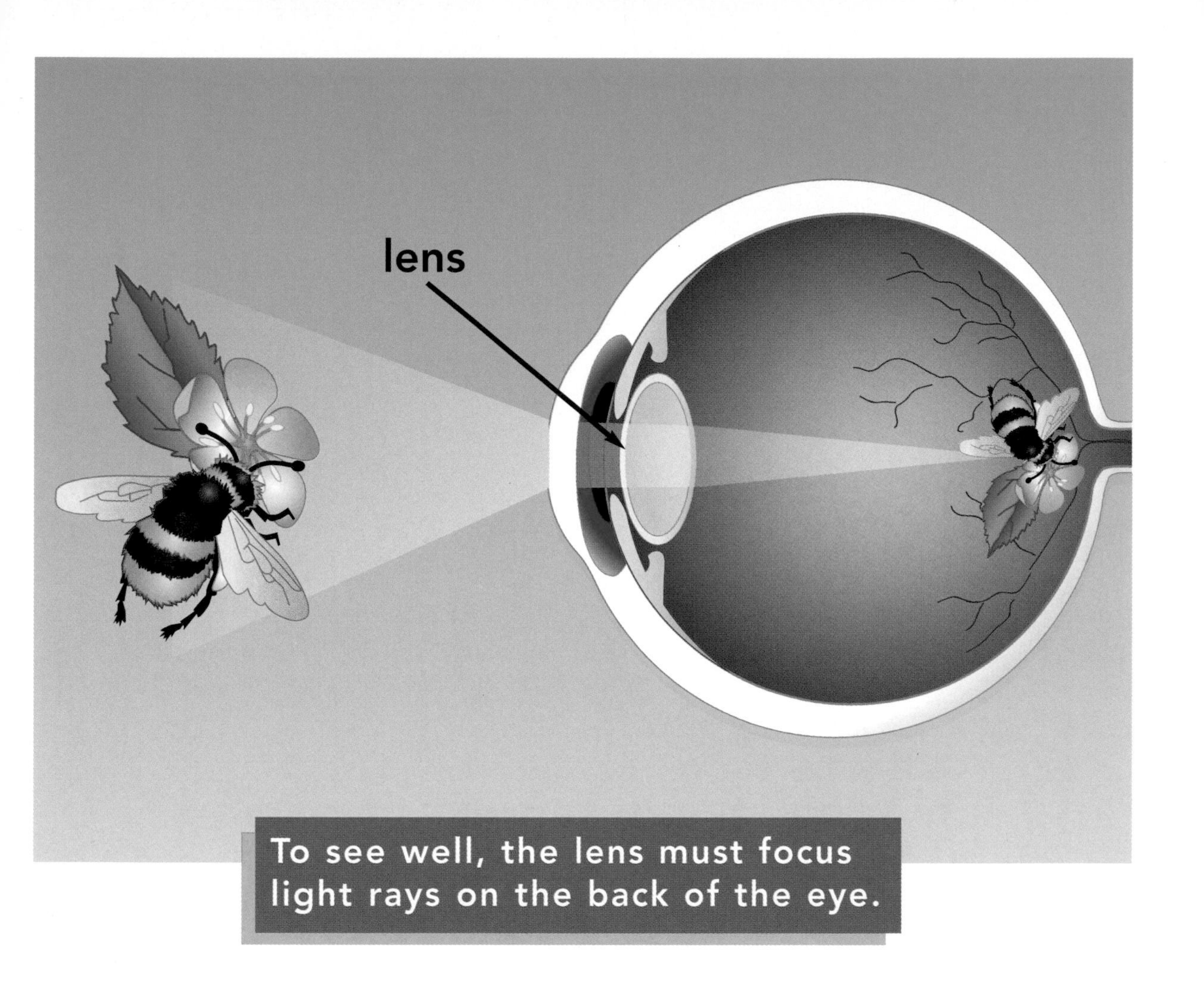

To see well, the lens must focus light rays on the back of the eye.

with 20/20 vision, the lens bends the rays so that they come to a point on a spot at the back of the eye.

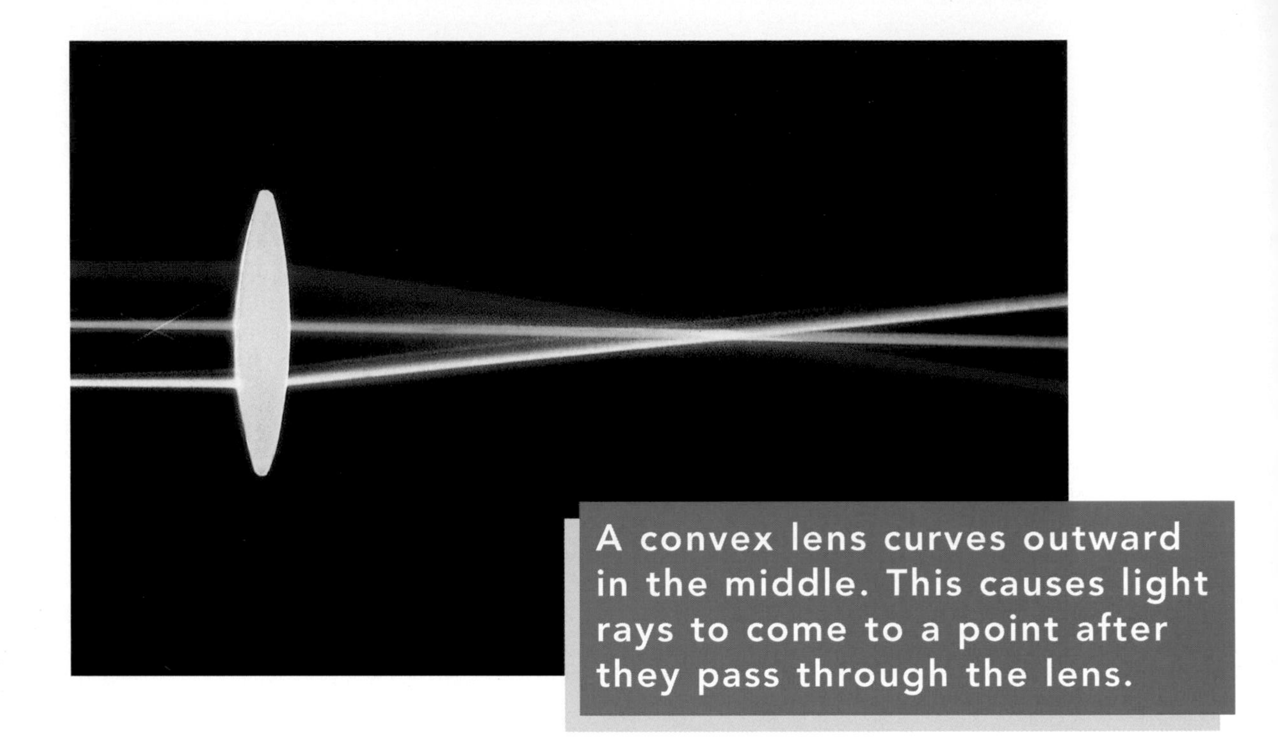

A convex lens curves outward in the middle. This causes light rays to come to a point after they pass through the lens.

The lens in the eye is convex. A convex lens bends light rays so that they come together at a single point. Convex lenses are used to make eyeglasses for people who have trouble seeing objects that are near.

Some lenses bend light rays so that they spread apart. This type of lens is called a concave lens. Concave lenses are used to make eyeglasses for people who have trouble seeing objects that are far away.

A concave lens curves inward in the middle. This causes light rays to spread apart after they pass through the lens.

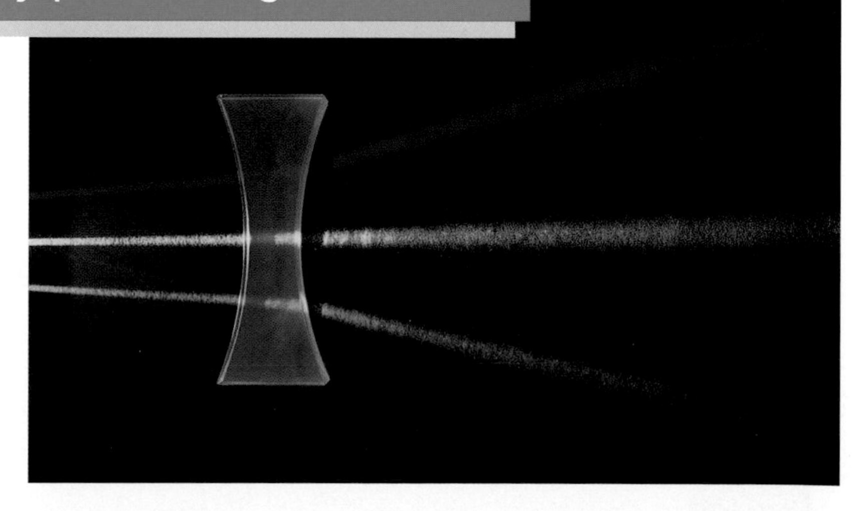

Because it magnifies, a hand lens is also called a magnifying glass.

Convex lenses can be used to examine objects that are too small to see with just the eyes. For example, a hand lens contains a convex lens. A hand lens magnifies, or makes something look bigger. Did you know that water can magnify just like a hand lens?

Using Water to Magnify

You will need:
- small round fishbowl or clear gallon-size glass jar
- marker
- white paper

Fill the fishbowl or jar with water. Print your name on the white paper. Hold the paper close to one side of the bowl or jar. Look through the water at the paper from the opposite side. How does your name look?

The letters you see may be a little fuzzy and bent. This happens because of the way light rays are bent as they pass through the glass and water.

The letters you printed seem larger when you look at them through the water. The water in the glass acts as a lens. Light rays bounce off the white paper and pass through the water. As they pass through the water, the light rays are bent. They spread out when they pass through the other side of the bowl or jar. This makes the letters appear larger, just as they would if you used a hand lens to view them.

Use your fishbowl or jar magnifier to examine various household objects. Items you can look at include a coin, paper clip, ruler, and pencil. All these objects should appear larger. Is there some way to tell how much larger a lens makes something appear?

Zachary

Checking the Magnification

> *You will need:*
> - marker
> - white paper
> - ruler
> - hand lens

Print your name on a sheet of paper. Use the ruler to make sure that each letter is 1 inch (2.5 centimeters) high. Use your hand lens to examine your name. Move the lens until the letters are clear. What you see through a hand lens is called an image. As you look through the lens, copy the letters on a sheet of paper just as they appear in the image.

> Print the letters the same size as they appear in your lens. You may not be able to see all the letters of your name through the lens at one time.

You are now ready to find the magnification of your hand lens. The magnification tells you how many times larger a lens makes something appear. Measure the height of the letters you drew from the image. Remember that the letters you first drew are only one inch high. Suppose the letters you drew from the image are 2 in. (5 cm) high. In this case, the letters in the image are two times larger than the real letters. The magnification of this lens would be 2 times. If the letters in the image are 3 1/2 in. (8.9 cm) high, then the magnification of the lens is 3 1/2 times. What is the magnification of your hand lens? Does your hand lens do anything else besides magnify?

This lens magnifies objects, such as letters, to appear twice their actual size. The magnification of this lens is 2 times.

Turning Things Around

You will need:
- source of light such as a sunny window or television set
- hand lens
- white paper

On a sunny day, turn off all the lights in a room with a window. You can also do this experiment at night. In this case, turn on a television to use as a source of light. Stand a few feet away from the window or television. Hold your lens so that the light shines through it. If you use a window, do not use the hand lens to look at the sun because the sunlight will damage your eyes. Never use a hand lens to focus light rays from the sun because doing that can start a fire.

Ask someone to hold the sheet of paper so that the light shines through the lens and onto the paper. Start by holding the lens close to the paper. Slowly move the lens away from the paper and toward the light source. Keep your eyes on the paper.

You may not see an image on the paper right away. If you don't, try standing closer to the light source or moving the lens farther from the paper.

When the lens is just the right distance from the paper, you should see an image of what is in the window or what is on the television screen.

How does the image on the paper compare to the real thing? Notice that the image is upside down and reversed.

What Can You See?

To understand how nature works, scientists must observe things very closely. To do this, scientists often use their senses. They look, hear, touch, smell, or even taste things that other people might ignore. At times, however, even their senses are not enough to explore nature.

Lenses are used to make both microscopes (left) and telescopes (above).

In these cases, scientists need to use instruments such as microscopes and telescopes.

You can work and think just like a scientist. Many times your senses will be enough to observe things. Sometimes, however, you will need to use a hand lens. A hand lens will give you a closer look at the world around you. With a hand lens, you will see things that you cannot see with just your eyes. For example, use a

If your hand lens is powerful enough, you should see the state names on the Lincoln Memorial.

hand lens to look at a five-dollar bill. How many state names can you see on the Lincoln Memorial shown on the back of the bill?

A five-dollar bill is only one of many things that you can examine with a hand lens. In fact, you can use a hand lens to examine everything around you. With so much to examine, where should you look first? Perhaps the best place to start looking is not around you, but on you. The first thing to examine with a hand lens might be your hand.

Examining Your Hand

You will need:
- hand lens
- lamp
- sharp scissors or nail clipper

Hold your hand under the lamp. Use the hand lens to examine the skin on the back of your hand. The skin may look smooth to your eyes. Under a hand lens, however, you can see all the tiny grooves that criss-cross to form different patterns on your skin.

You may also see tiny hairs on the back of your hand.

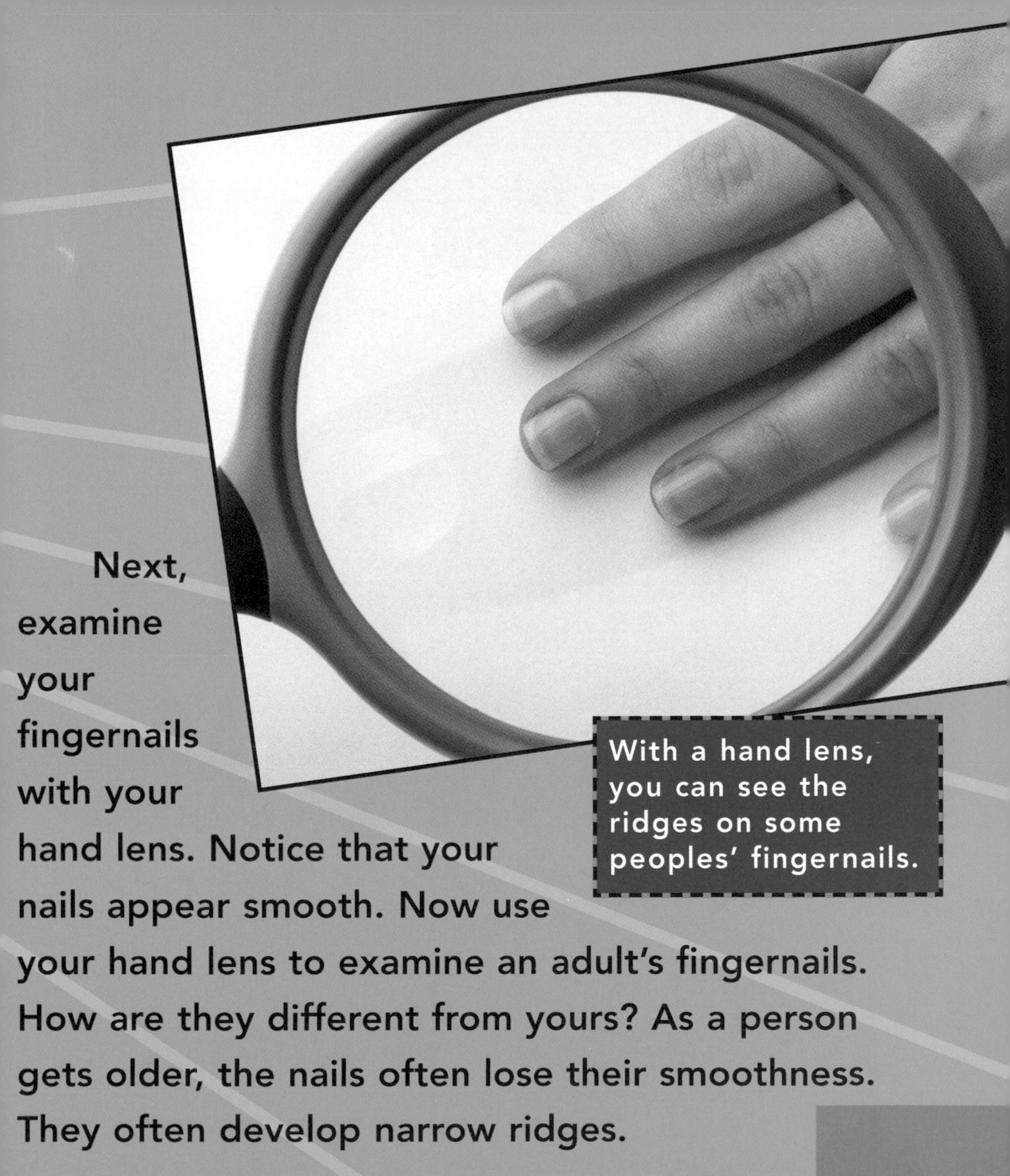

With a hand lens, you can see the ridges on some peoples' fingernails.

Next, examine your fingernails with your hand lens. Notice that your nails appear smooth. Now use your hand lens to examine an adult's fingernails. How are they different from yours? As a person gets older, the nails often lose their smoothness. They often develop narrow ridges.

Trim a small piece of your fingernail with sharp scissors or a nail clipper. Cut this piece into several smaller pieces. Examine one of these smaller pieces with your hand lens. Can you see the three layers that make up a fingernail?

Now, examine the white strip at the base of your nail. This is the cuticle. The cuticle acts as a seal to prevent germs from entering the body through the space between the nail and the skin. Cutting the cuticle is not a good idea. Without a cuticle, germs can enter the body and cause an infection. Besides taking a closer look at your hand, what else can you learn about yourself with a hand lens?

Examining Your Tongue

You will need:
- mirror
- hand lens
- two small drinking glasses
- teaspoon
- sugar
- cotton swabs

Look in the mirror and stick out your tongue. Use a hand lens to examine the surface of your tongue by looking at your reflection in the mirror. Like the skin on your hand, the surface of your tongue is not smooth. It is covered

If you have trouble examining your own tongue, look at someone else's with your hand lens.

Do not touch too far back on your tongue. If you do, you might gag.

with tiny bumps. Use your hand lens to find out where these bumps are located on your tongue. Can you see that they are found all over it? These tiny bumps allow you to taste things such as sugar.

Fill both glasses with water. Add two teaspoons of sugar to one glass. Stir until the sugar is dissolved. Dip a cotton swab into the sugar water. While looking in the mirror, use your hand lens to touch the cotton swab to the tiny bumps near the tip of your tongue. Do you taste the sugar?

Rinse your mouth with plain water. Now, use the cotton swab to touch the tiny bumps along the sides of your tongue. Rinse again. Next, touch the cotton swab to the tiny bumps in the middle of your tongue. Rinse again. Finally, check out the tiny bumps near the back of your tongue. Where do you taste the sugar?

Nature is full of things you can explore with a hand lens, including plants such as this fiddlehead fern.

Your hand and tongue are only two of the large number of things that you can examine with a hand lens. Use your hand lens to explore whatever you can find. Use it to look at a feather, a flower, a fly, or a fiddlehead fern. No matter what you look at, you will see more than what you can see with just your eyes. If you look closely enough, you can also tell a lot with a hand lens.

What Can You Tell?

The famous detective Sherlock Holmes used a hand lens to examine the scene of a crime. Looking closely at a footprint left in mud, he could tell the person's height and weight. Looking at ashes, he could tell if they came from a pipe or cigar.

Imagine that you are looking at a person's fingernails with a hand lens. Suppose you could see only the person's fingernails and nothing else. Could you tell if the person were young or old? Remember that as a person gets older, ridges appear on the fingernails. So you might be able to tell if a person is young or old just by examining his or her fingernails with a hand lens. What else can you tell by using a hand lens?

Uncovering an Identity

You will need:
- white paper
- soft lead pencil (#2 or softer)
- clear sticky tape
- 5 x 7 white index cards
- soap
- hand lens

Spread out your fingers and place them on a piece of white paper. Use a pencil to draw an outline around each of your fingertips. Remove your hand from the paper. Rub the sharpened end of the pencil across each outline on the paper. Next cut five pieces of sticky tape, each about one inch long.

Gently press your thumb on the paper so that it picks up some of the rubbings. Gently stick a piece of tape onto your darkened thumb. Remove the

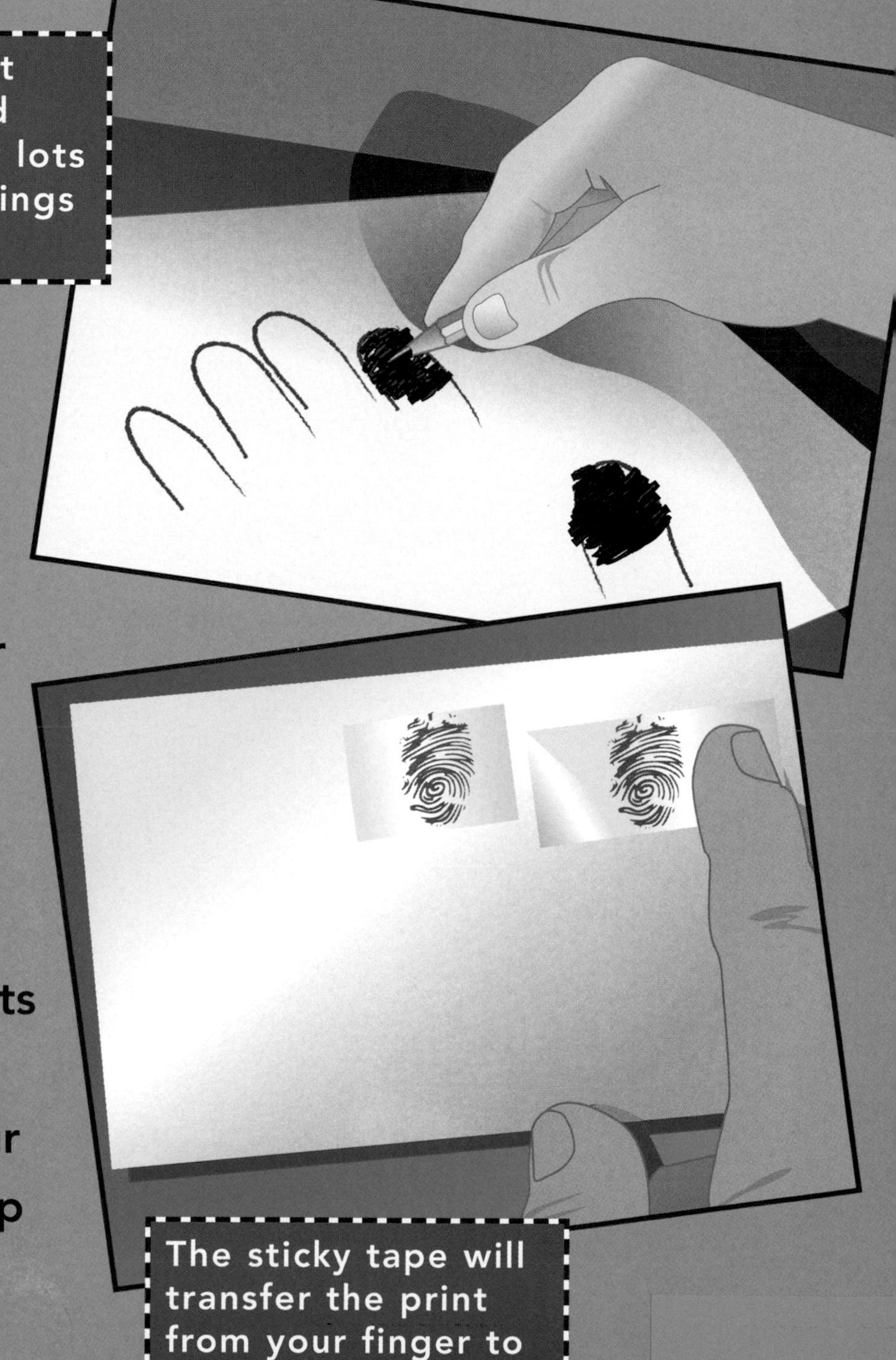

Make sure that you press hard enough to get lots of pencil rubbings on the paper.

tape and place it on an index card. Do the same with your other four fingers. You should now have a set of your fingerprints on the index card. Wash your hands with soap and water.

The sticky tape will transfer the print from your finger to the index card.

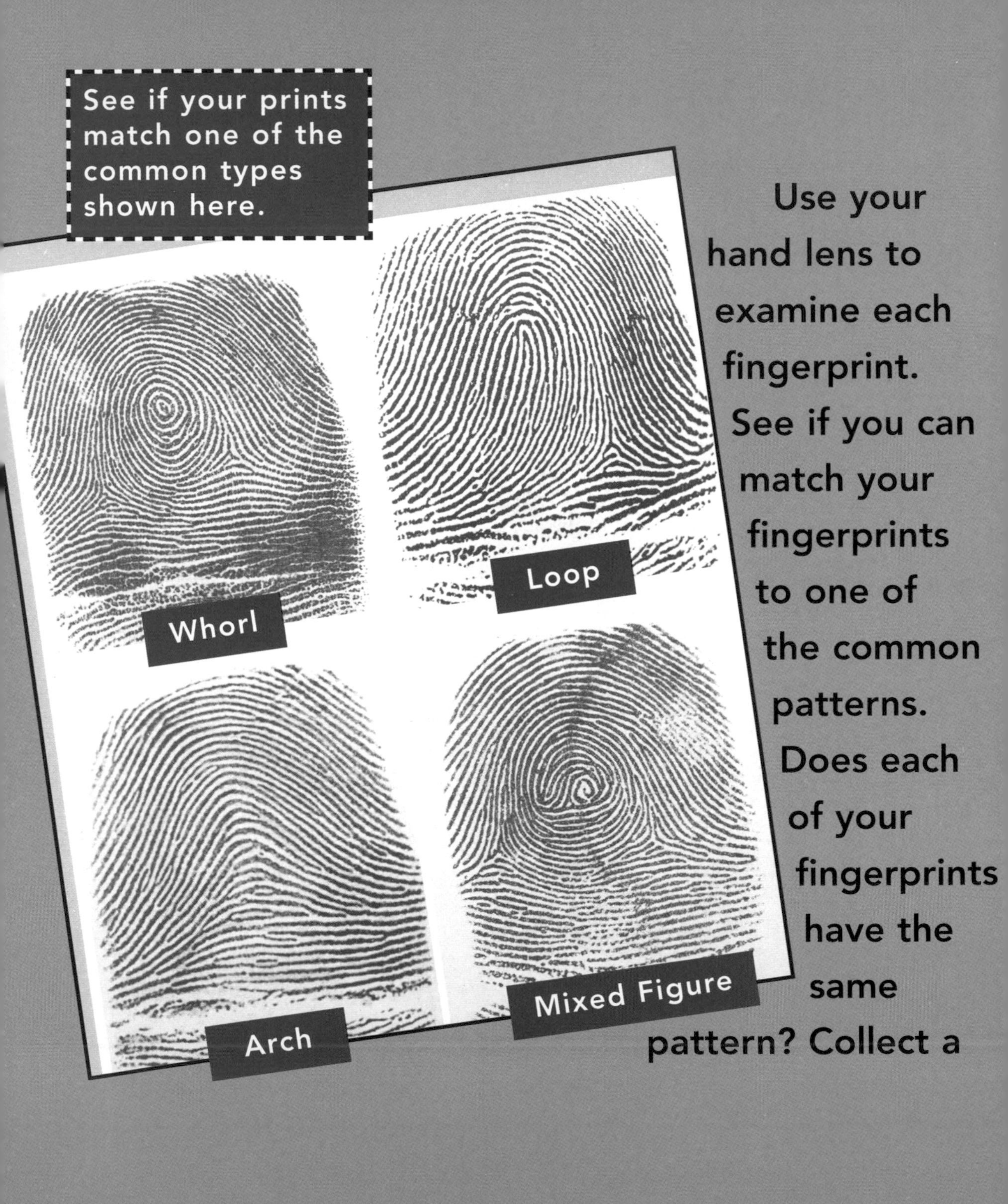

See if your prints match one of the common types shown here.

Use your hand lens to examine each fingerprint. See if you can match your fingerprints to one of the common patterns. Does each of your fingerprints have the same pattern? Collect a set of fingerprints from each member of your immediate family. Be sure to write the name of each person on the back of the index card.

No two people have the same fingerprints, not even identical twins. Have a family member give you a fingerprint without telling you whose it is. See if you can tell whose fingerprint you are examining. Do this by using a hand lens to compare the unknown fingerprint with those you have on record.

Save all your fingerprints for a later experiment. But first, let's learn what else you can tell about your family with a hand lens.

Identifying the Real Thing

You will need:
- several family members or friends
- pencils
- index cards
- hand lens

Have one family member or a friend sign his or her name on an index card. Have the person write "original" on the card. Give this card to several of your family members or friends and ask each person to copy the name on a separate index card just as it is written on the original. The person who wrote the original should write his or her name on another card. When everyone is finished, collect all the cards, including the one marked original.

Examine all the signatures with a hand lens. To tell if the signature is a copy, carefully compare each letter to the original. Look for places where the person may have gone over a letter to try to make it look more like the original. Look to see if the letters are slanted the same way. Can you tell which signature was written by the person who wrote the original signature?

Notice how the letters in these two signatures do not slant the same way. These two signatures were not written by the same person.

Fun With a Hand Lens

You have learned that a hand lens bends the light rays that pass through it. By bending light rays, a lens can make something seem bigger. With a hand lens, you can see things that you cannot see with your eyes, even if you have 20/20 vision.

Searching for Prints

You will need:
- dark surface
- talcum powder
- hand lens
- piece of paper
- clean, dry brush
- clear sticky tape
- 5 x 7 white index card

Find a dark surface in your home that people often touch. The surface could be a counter top, the surface of a desk, or the shiny cover of a book. Try to find a fingerprint on this surface.

A fingerprint is often left by perspiration whenever a fingertip touches a surface.

Police always search for fingerprints that may have been left at the scene of a crime.

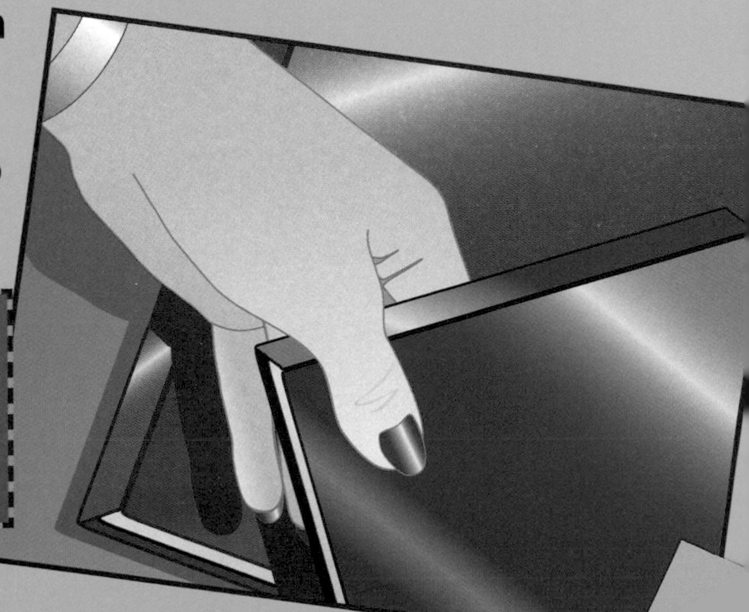

To find a print, gently sprinkle some talcum powder over the dark surface. Use the hand lens to search for something that looks like a fingerprint. When you do, sprinkle some powder on the paper. Dip the brush into the powder on the paper. Lightly dust any area where you think you see a fingerprint. If you find a fingerprint, gently brush away the excess powder.

Unroll about 6 in. (15.2 cm) of tape. Place one end of the tape to the right of the fingerprint. Carefully lower the tape so that it covers the fingerprint. Leave one end of the tape free. Smooth out the tape to force out the air bubbles. Lift the tape by the free end. Place the

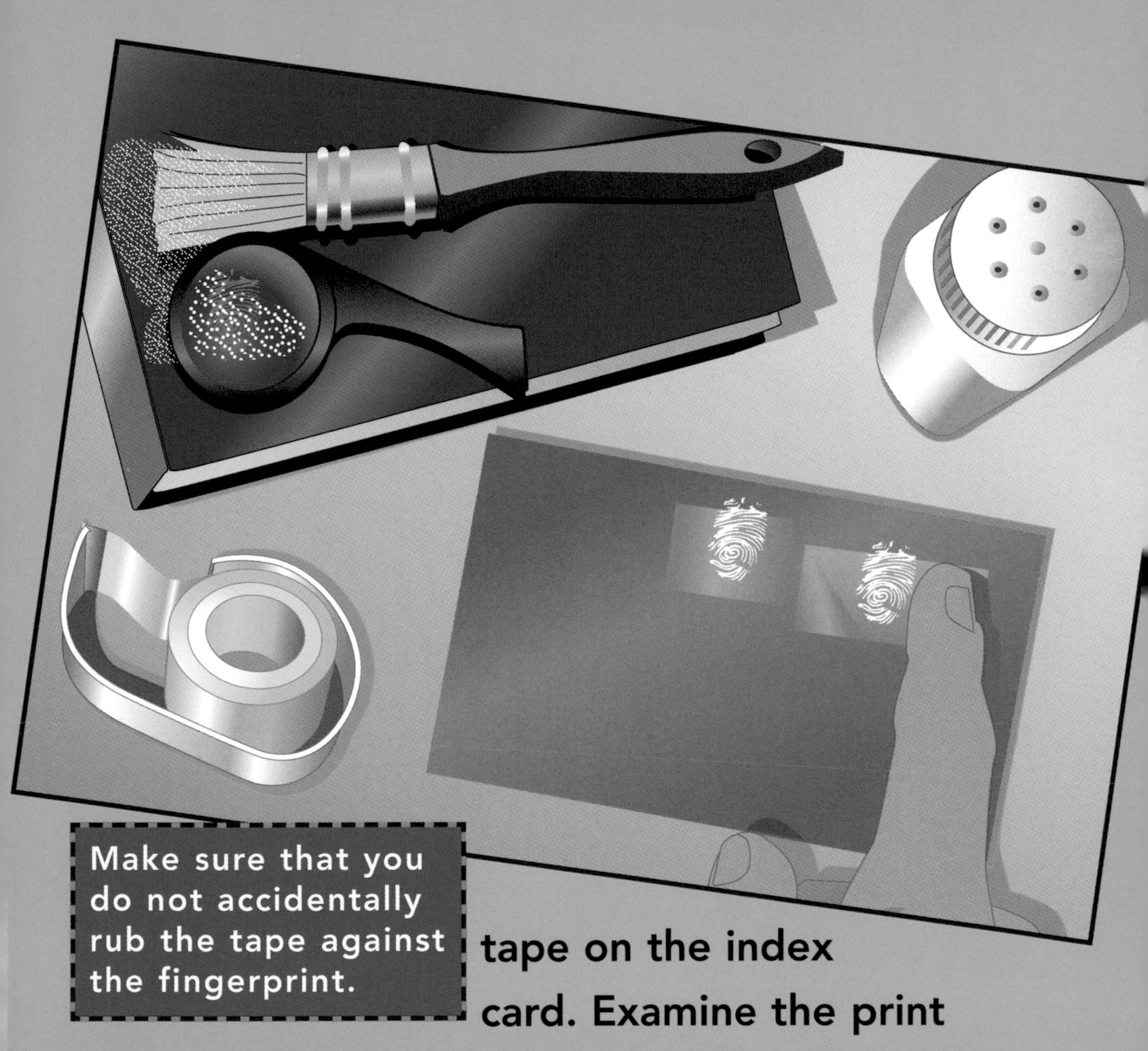

Make sure that you do not accidentally rub the tape against the fingerprint.

tape on the index card. Examine the print with a hand lens. Can you identify the print by comparing it to the ones you saved from your previous experiment with fingerprints?

To Find Out More

If you would like to learn more about hand lenses, check out these additional resources.

Rainis, Kenneth G. **Exploring with a Magnifying Glass.** Franklin Watts, 1991.

Ross, Michael Elsohn. **The World of Small: Nature Explorations with a Hand Lens.** Yosemite Association, 1993.

VanCleave, Janice. **Micro-scopes and Magnifying Lenses: Mind-Boggling Chemistry and Biology Experiments You Can Turn into Science Fair Projects.** John Wiley & Sons, 1993.

Organizations and Online Sites

Document Exam
http://www.documentexpert.com/question.htm

This site explores how to tell if handwriting has been copied or forged.

Edmund Scientific Corporation
60 Pearce Avenue
Tonawanda, NY 14150-6711
800-728-6999
http://www.edmundscientific.com

You can purchase an Optics Discovery Kit to learn more about lenses, mirrors, and telescopes. You can also explore light in more detail, including color and holograms.

Museum of Science and Industry
57th Street and
Lake Shore Drive
Chicago, IL 60637
http://www.msichicago.org/ed/mystery/fprint.html

This site shows you the four basic fingerprint patterns and then challenges you to identify four unknown patterns.

Optical Society of America
2010 Massachusetts Ave., NW
Washington, DC 20036
202-223-8130
http://www.osa.org/

Check out this site if you have a question about optics, which is the study of the eye and how we see.

Optics for Kids
http://www.opticalres.com/kidoptx.html

This site covers the basics of optics and has links to several other useful sites.

Important Words

concave lens lens that curves inward in the middle

convex lens lens that curves outward in the middle

inverted flipped upside down and backwards

lens anything that bends light rays as they pass through it

magnification how many times bigger something is made to look

magnify to make something look bigger

Index

Meet the Author

Salvatore Tocci is a science writer who lives in East Hampton, New York, with his wife, Patti. He was a high school biology and chemistry teacher for almost thirty years. As a teacher, he always encouraged his students to perform experiments in order to learn and understand science. He depends on a hand lens to see small details whenever he is working on his HO-scale train layout.

Photographs © 2002: Digital Vision: 1; Fundamental Photos, New York/Richard Megna: 11; James Levin: cover, 4, 12, 24, 25, 27, 28; Photo Researchers, NY: 30 (Michael P. Gadomski), 10 (David Parker/SPL); Photodisc, Inc.: 2: The Image Works/Eastcott-Momatiuk: 7; Visuals Unlimited: 20 left (Steve Callahan), 20 right (Larry Stepanowicz); www.fingerprints.demon.nl: 36.

Illustrations by Patricia Rasch